Getting Started

Clicker Training
for Dogs

for Dogs

by Karen Pryor

A
Karen Pryor

Clicker
Book

Getting Started

Clicker Training
for Dogs

by Karen Pryor

New Expanded Edition!

Sunshine Books, Inc.
49 River Street, Suite 3
Waltham, MA 02453-8345
www.clickertraining.com

Other titles in this series

Getting Started: Clicker Training for Cats
by Karen Pryor

Getting Started: Clicker Training for Horses
by Alexandra Kurland

Getting Started: Clicker Training for Dogs, Revised Edition
© 2002 by Karen Pryor

For information contact:
Sunshine Books, Inc.
49 River Street, Suite #3
Waltham, MA 02453
781-398-0754
www.clickertraining.com

Library of Congress Control Number: 2002092015

ISBN 1-890948-11-X
Book design by Codesign, Boston

Printed in the United States of America
10   9   8   7   6   5   4   3   2   1

Contents

Introduction

Dear Dog-Lover,

Welcome to clicker-training. Clicker training is not someone's new gimmick or special method. Clicker training is dog trainers' slang for a positive reinforcement training system based on operant conditioning, a set of scientific principles describing the development of behavior in which the animal "operates" on the environment, instead of the other way around.

Operant conditioning goes to the root of how animals learn in the natural world; therefore the principles can be applied in any circumstance. The possibilities for dog training are tremendous. Clicker trainers are developing ways to use this new technology with police patrol dogs; with service and hearing dogs; with puppies, even before weaning; for competition obedience and agility; in hunting, field trials, and tracking; in the breed show ring; in "good manners" classes for pet owners; and in the home.

About Clicker Training

Clicker training does not really depend on the clicker or on food. Clicker training depends on reinforcers, which may be anything the dog likes (toys, petting, etc.) presented with a correctly timed signal in such a way that information is communicated between trainer and pet. We start with food as the reinforcer and the clicker as the signal because these are excellent teaching tools, for both you and your dog. Together you and the dog will be learning a new way of interacting.

Anything you need your dog to know can be taught by reinforcement training. After the dog has learned to understand what behavior you want, how to do it, and when to do it, you can replace the click with a word, and the food with a pat; a word and a pat, you always have with you.

Once your dog acquires a clicker-trained behavior, unless you add new rules the dog will have that behavior for a lifetime. When the dog knows several behaviors, you no longer will need to click or to praise each one; you can reinforce once, for a whole repertoire. Eventually, you'll find that you get out your clicker only to sharpen up a response or to teach the dog something new—or just to have fun with your dog: clicker-training is fun for both of you.

But how do you get started? An experienced clicker-trainer could teach you the opening moves in a few minutes; but clicker-training teachers are in short supply, and many of us have to learn on our own. This little book is meant to get you going: to give you the skills to develop your first clicker-trained behaviors. When you have personally accomplished even a little clicker-training, you will find that other books and videos on clicker training will make more and more sense to you. Many thousands of clicker trainers have begun successfully with the information in this book.

The Resources section of this book will tell you where to look for additional information. Do make use of these resources, some of which, like the Internet clicker lists, are free. Also, as a new clicker trainer you may find that you are becoming a resource, yourself, to other pet owners and dog trainers around you. Enjoy it! This is a creative field, and all of us can make a contribution.

– Karen Pryor

A Dog and a Dolphin

Dogs, Dolphins, and Training

If you've seen trained dolphin shows at oceanariums or on TV, you will know that dolphins appear to be wonderfully trainable. On command they exhibit all kinds of precision behavior, including splendid acrobatics and interactive behavior with other dolphins or with human swimmers. The audience marvels at how eagerly they respond, and how intelligent they must be; wouldn't it be nice if dogs responded like that?

As we dolphin trainers know well, the truth is that dolphins aren't geniuses, and neither are dolphin trainers. The dolphins' speed, precision, and obvious enjoyment of their work is due entirely to the principles dolphin trainers use in training them. And the same techniques can be used on dogs.

Omitting Punishment from the Start

The first thing to understand about dolphin training is that we are working with animals you can't punish. No matter how mad you get—even if the animal makes you mad on purpose, by splashing you from head to foot, say—you can't retaliate. You can't use a leash or a whip or even your fist on an animal that just swims away. You can't starve a dolphin into being cooperative. Dolphins get their fresh water from the fish they eat; if you take away the fish, they rapidly become dehydrated, they lose their appetites altogether, and then they die. Finally, you can't even yell at a dolphin, because they don't care.

Maybe you're thinking "I bet I could think up a way to punish a dolphin...." and I bet you could; but it doesn't matter, because dolphin trainers don't need it. Trainers can get whatever they want from a dolphin, using positive reinforcers only: mostly just a chirp or two from a training whistle, and a bucket of fish. We "shape" every behavior by positive reinforcement. We use positive reinforcement to elicit prompt and correct response to commands—to achieve obedience. We can even use positive reinforcement to discipline an animal—to control misbehavior such as attacking a tank mate or refusing to go through a gate. This sophisticated use of positive reinforcement results in an animal that works brilliantly and loves to work.

The methods we use to train dogs often include the use of force, both to put the dog through required movements and to correct the dog when it makes mistakes, which it inevitably does. Although we may also use praise and petting, unavoidably the dog experiences some confusion, fear, and maybe even physical pain in the training process. Some dogs tolerate these negative experiences well, but dolphins, being wild animals, would not. If you were to train a dolphin by these techniques, the dolphin might learn, but it would offer a sluggish, sulky, unreliable performance; and it might well begin to exhibit aggression toward people. (Does that sound like any dogs you know?)

On the other hand if you train a dog the way we train dolphins, through positive reinforcement, the dog behaves just like a dolphin: it becomes eager, attentive, precise, cooperative, and capable of fantastic performance. Here's how it's done.

The Magic Signal: Conditioned Reinforcers

When I talk to dog trainers a big misconception I run into is that positive reinforcement just means "food." Wrong. The crucial element in getting wonderful behavior out of a dolphin is not the food reward. The dolphin is not working for the fish; the dolphin is working for the whistle. The sound of the whistle is the magic signal that brings about that wonderful performance.

The first step in training a dolphin is to teach it that every time it hears a whistle, it's going to get a fish. Once the animal knows that the whistle means "Fish is coming" the trainer can use the whistle to mark a behavior she likes, and then, gradually, to shape or develop something more complex, such as a response to a cue.

For example: Suppose, on several occasions, the dolphin heard the whistle (and later got a fish) when it happened to be jumping in the air. Soon it would start jumping every time the trainer showed up. Then it might be allowed to discover that jumping only 'works' when the trainer's arm is raised. So a raised arm becomes the green light, as it were, for jumping.

The trainer could gradually impose other conditions — jumping only 'works' when the direction of the jump is away from the trainer and toward the audience; when the jump is higher than four feet; when the jump occurs within three seconds after the arm is raised. At the end of a few training sessions the trainer has trained the dolphin to 'take a bow,' on command and with precision; and the dolphin has trained the trainer too: "All I have to do is make a certain kind of jump when she sticks her hand up, and she immediately gives me a whistle and a fish every time!"

Note that the whistle is not used as a command. It does not tell the dolphin to start doing something—the hand signal does that. The whistle tells the dolphin, during or at the end of a behavior, that the trainer likes that behavior and the dolphin deserves a fish for it. (You don't have to stick with food, either; you can also associate a conditioned reinforcer with a pat, or a toy, or maybe just another chance to work).

The whistle has now become a conditioned reinforcer. In the language of psychologists, food, petting, or any other pleasure is an unconditioned reinforcer—something the animal would want, even without training; the whistle, a conditioned reinforcer, is something the animal has learned to want. (Some people use the term 'primary reinforcer' for food and 'secondary reinforcer' for the signal; I avoid those terms because I find it leads people to think that if the whistle is 'secondary' it should occur after the food, which of course makes it meaningless to the animal and useless as a training tool.)

Why the Conditioned Reinforcer Is Crucial

What would happen if you tried to train a dolphin to do a simple jump, away from you, on cue—without the whistle? First, you could not possibly time the fish to arrive when the animal was in mid-jump; so no matter what kind of jump the animal gave, it would either get the fish later, or get no fish at all. It would have no way of telling why you rewarded one jump over another, or what you liked about the jump. Was it the height? Or maybe the way the animal took off or landed? To develop a jump of a particular height, timing, and direction you would have to eliminate mistakes by trial and error over many, many repetitions; you would be lucky if the animal didn't get bored (and the trainer too!) before the performance was correct and reliable.

Because of this lack of information, the trainer who uses food reward without using a conditioned reinforcer first typically produces an animal that works eagerly (as long as it is hungry) but learns slowly. We see this in dogs that have been rewarded with lots of treats without any clear signal as to exactly what behavior earned the treat; they often seem to be enthusiastic and friendly but they don't know anything.

Also when a trainer uses food without a marker signal the animal is apt to look toward the trainer for food all the time. Horses nose your pockets and dogs lick your hands. Dolphins hang around the training station and worship the fish bucket. And with the animal constantly looking at the trainer it would be difficult to train our dolphin to jump facing away from the trainer, toward the audience.

Once you've established the marker signal, however, you can use the whistle to identify behavior that occurs at a distance, or with the animal facing away from you, with no trouble at all. And the well-conditioned animal, instead of nosing around for a snack, is going on about its business, but also attentively listening for the magic sound, whatever else it may be doing: in horses and dogs as well that attentiveness is a valuable training asset in itself.

Because of the split-second timing that the conditioned reinforcer makes possible, the whistle also communicates just exactly what it is that the trainer is looking for. This allows you to teach the animal what you want, in a very clear way, one detail at a time. For example, let's say that a dolphin has assimilated one rule ("Jump facing this way") and you know that because the animal almost always jumps with the proper orientation when you signal it to jump. Now you can add another detail or rule. You decide "I'll only reinforce the higher jumps." Pretty soon the dolphin has learned one more detail ("I have to jump facing this way and jump this high").

This step-by-step process may seem elaborate but in practice it is a fantastic short cut to complex trained behavior. Even with a naive dolphin, a trainer can develop an on-cue, spectacular, and very specific behavior, such as the bow I've described, in two or three days—sometimes, if things go well, in a single ten-minute training session. Many times in my dolphin training experience I have "captured" a behavior, shaped it into something special, and put it on cue in a single training session—and so have other dolphin trainers.

How about Dogs?

You can easily experience dolphin-training your own dog, using a conditioned reinforcer, in one quick ten-minute experiment. Some dogs are afraid of whistles. A handy conditioned reinforcer for dogs is a clicker, a child's toy that goes click-click when you pinch it; they are available in toy and novelty shops and many Web sites. You may substitute a bottle lid, a pocket stapler, or a retractable pen for your signal.

Get yourself a clicker and a few treats. Make the treats small enough so that you can give the dog fifteen or twenty treats without him filling up. Some dogs will work for kibble, especially just before dinnertime, but you might have to go to something more tempting; in demonstrating clicker training with inexperienced dogs I generally use diced chicken. Teach the dog the meaning of the click by clicking the clicker and giving a treat, four or five times, in different parts of the room or yard (so the dog doesn't get any funny ideas that this only works in one place).

Then click the clicker and delay the treat a few seconds; if you see the dog startle and actively look for the treat, you will know the signal has become a conditioned reinforcer. Now you can establish a behavior—we call this "shaping."

An easy behavior to shape is "Chase your tail." There are, of course, as many ways to elicit this behavior as there are trainers to think them up: you could turn the dog around by its collar; you could put bacon grease on its tail tip so the dog circles to lick its tail. Here's one way to shape the behavior "from scratch" without any prompting.

Stop clicking and just wait. Your dog may be intrigued and excited by now; when you do nothing, the dog is likely to move around, and maybe even to whine and bark. The instant the dog happens to move or turn to the right, click your clicker. Give the treat. The click is also information for you; it teaches you good timing because you can tell if you clicked too soon or too late, a distinction that is much harder to make if you are using a spoken word.

Wait again. Ignore everything the dog does, except moving to the right (don't hold out for miracles; one turn of the head or one sideways step with the right front paw is all you need.) If you "catch" the behavior—if your timing is good—in three or four reinforcements you will see your dog turning to the right further and more often.

Now you will find you don't need to reinforce just a single step to the right, but can reinforce right turns that go several steps, perhaps through a quarter of a circle, and from a quarter turn, a full circle may come very quickly.

That's a good time to stop the first session: quit while you're ahead, is the golden rule. Put the clicker away, with lots of hugs and praise, and try to again the next day, starting with a single step, then a quarter circle, and then more. It will come much faster the second time.

From one circle, the next step is to get two circles, and the next step—an important one—is to go for a variety, rewarding half a circle sometimes, then two circles, or one, or three full turns, or just one and a quarter; this keeps the dog guessing. The click might come after one turn, or two, the dog doesn't know, so he keeps turning, faster and faster; and thus you begin to develop an amusing whirl after his own tail.

This is a silly trick, of course, and not very dignified. There are other behaviors you could use for practice, such as targeting, in which you shape the animal to touch some object with its nose (sea lion trainers teach their animals to "target" on the trainer's closed fist; then by holding their fist on the ground or in the air or over a stand they can move the sea lion where they want, without using force). The purpose of this experiment is not to teach the dog the trick, but to show you how to use a conditioned reinforcer to shape behavior, and how effective this kind of reinforcement can be.

Why do you need to use a clicker? Why couldn't you just use your voice, and the words "Good boy," as the conditioned reinforcer? One reason is that you can't say a word, even "Good boy," with the split-second precision that you can achieve with a click. With the clicker and a little practice you can reinforce very tiny movements—one paw stepping to the right—in the instant that it occurs; a praise word is inevitably rather 'fuzzy,' because it takes longer.

The second difficulty with using a word is that we also talk near our dogs and even to our dogs when we are not reinforcing them. It is hard for the dog to sort the meaningful words out from the stream of noise we make; but the clicker is unlike any other sound in the room, and its meaning is crystal clear. You will in fact see the difference very clearly in the way the conditioned dog responds to the clicker (electric attention, galvanized, thrilled) as compared to the way the dog responds to "Good dog!" (Hunh? Oh. Smile, wag.)

Using a Conditioned Reinforcer in the Real World

Well, I've heard dog trainers say, the cricket's good for tricks but not for anything else—you can't, for example, use it in the obedience ring. Of course not, and you don't need to; the clicker's value is in shaping new behavior, or refining details. It's not necessary in exhibiting behavior the animal has already learned. But even in the accomplished champion working dog, the conditioned reinforcer can be a useful training tool. One competitor told me he taught his Doberman to understand the clicker, and then used it to reinforce her for looking into his face, instead of away from him, while she worked. "It's as if she were grateful for the information, really. It cleared up the vagueness, for her," he said. Of course once the dog had come to understand what was wanted, she did it correctly in the ring, without any clicks.

Don't think, however, that people never use a conditioned reinforcer in the ring; all the trainer has to do is establish a signal the dog is aware of that no one else notices. I know a keen obedience trainer who uses a barely audible sniff as a conditioned reinforcer. I have seen a competitor convey "Great job!" (as evinced by the overjoyed expression on the dog's face) just by touching one finger to her dog's head. One competitor I know has taught her dog, Rex, that treats are called 'Billy.' Then as the dog performs in the obedience ring, she can reinforce an especially good behavior— a nice recall perhaps—with what appears to be a command: "Billy, heel!" No one questions why she doesn't use the dog's usual name in the ring.

Once a behavior is learned, using a conditioned reinforcer allows you not only to delay the food, without loss to the performance, but to give less food overall; you don't have to worry that your animal will fill up before the job is done. One example: at dog shows I have often noticed handlers repeatedly baiting or feeding a dog to get nice pose or alert look. Whenever I see that food-food-food going down the dog's throat I know at once that the person doesn't understand conditioned reinforcers! How much more effective it would be to shape the pose, develop a verbal cue and then reinforce the dog with a click for assuming and holding the proper posture for a respectable length of time—with the actual food following later, outside the ring or when the judge has moved on.

The virtue of the conditioned reinforcer is that it works—it conveys information, and affects the animal's behavior—in all kinds of situations in which real reinforcement is not merely undesirable but in fact impossible. Think, for example, of how useful a simple conditioned reinforcer would be in training scent discriminations, tracking, long sits and downs, go-outs, pointing and flushing birds, and all other dog behaviors that require the animal to work away from you.

Controlling Misbehavior with Positive Reinforcement

It might seem unreasonable that you can control bad behavior with positive reinforcement instead of "correction," but dolphin trainers have many ways to do it. Here are three examples:

1 *Establish a conditioned negative reinforcer*
This does not need to be a signal that means "I'm going to beat you" (although you could establish that, too) but a signal that means "Nope, I'm not going to reinforce you." It tells the animal that some particular effort it is making is not going to pay off; the animal swiftly learns that whenever it gets this "red light" or "wrong" signal it should change what it is doing. You could use such a signal, for example, to help teach a dog not to jump up in greeting, but to keep its paws on the floor for a patting reinforcer.

2 *Use positive reinforcement to train an incompatible behavior*
Once in a dolphin show at an oceanarium, one animal took to harassing the swimmer who performed in the show. Rather than give the swimmer a stun gun (or some

such instrument of punishment) we trained the dolphin to push on an underwater lever, for a whistle and fish, and we asked the animal to do that when the swimmer was in the water. The dolphin could not press its lever and pester the swimmer at the same time; the behaviors are incompatible (and apparently lever-pressing was more reinforcing, because the swimmer harassment ceased). You are using this technique if you teach your dogs to lie in the living room doorway, at mealtimes, so they can't beg at the dinner table.

3 *The time-out*

Sometimes a dolphin does something really bad, such as showing aggression (swinging its head or teeth at the trainer's hand, for example). The instant this happens, you turn your back, snatch up your training props and fish bucket, and leave, for one full minute. That's the end of all the fun. The dolphin is apt to stick its head out of the water looking dismayed—"Hey, what'd I do?" In a few repetitions it learns to mind its manners. The doggy equivalent would be the same: put the treats and clicker away and end the fun.

Time-outs are used successfully by oceanarium trainers to eliminate aggression toward human swimmers, even in highly dominant animals such as adult male killer whales; the technique is, however, distressing to the animals and must be used sparingly.

Mental Attitudes

Using reinforcement is a lot of work for the trainer, because it forces you to think. Oh no, what pain! It's so much easier just to follow someone else's rules: If the dog makes a mess, rub his nose in it; if the dog doesn't heel, jerk the chain. However, in thinking out what you're going to reinforce, you'll be a better trainer. And the focus you will need, in order to perfect the timing of your reinforcers, makes training a thrill instead of a bore.

From the animal's standpoint, this kind of training is not a matter of learning how to stay out of trouble by doing what's required—a chore and nothing more. Instead, this kind of training gives the animal a chance to win, over and over, and also a chance to control at least part of its world. For example, from a dolphin's standpoint, once it has learned the meaning of the whistle, the training is not an exchange of commands and obedience, but a guessing game in which the dolphin tries to "discover" various ways to make the trainer blow that whistle. It is a game, with strict rules, but with equality on both sides. No wonder the dolphins enjoy their obedient trainers!

The effect of using the conditioned positive reinforcer is in fact far more powerful than merely giving free goodies could ever be. If you stop relying on control of misbehavior and start shaping good behavior with clear-cut conditioned signals for reinforcement, your dog will respect you in a new way; to your dog you will be making sense, at last.

Getting Started
A Few Easy Behaviors to Train with a Clicker

We usually think of a well-trained dog as a dog that doesn't do anything wrong: doesn't jump up on visitors, doesn't bark too much, doesn't get on the couch, and so on. We usually think of "training," therefore, as a matter of getting rid of bad behavior. Training is what we do to stop the dog from jumping up, barking, and pulling on the leash. So, conventional training often seems to consist largely of control and prevention: pushing on your dog or yanking on its leash to make it do something, and then correcting it, usually with another yank on the leash, when the dog does it "wrong," until he learns how to do it "right."

Clicker training, or operant conditioning, is quite different. Clicker training can produce a well-behaved dog, but by a new path. Instead of stopping the dog from doing the wrong things, we are going to teach it to do the right things: to greet visitors politely, to bark at the right time, to walk on a loose leash, to go only where dogs are supposed to go (on the floor, not the couch, on the lawn, not the flowerbeds) and so on.

Some people approach dog trainers as if they were going to a doctor, or a garage mechanic. "Here's my problem. Tell me what's wrong and how to fix it." Clicker trainers don't work that way. Our aim is to give you the tools for building behavior in general. When you understand the system, you can build any behavior you want.

There are no set recipes for developing specific behavior with the clicker and treats. I might train a specific behavior, such as walking politely on a loose leash, one way and clicker author Morgan Spector might train it another. Other clicker trainers might have good and workable ways of their own. What follows is a simple guide to a few basic behaviors, not to be followed religiously, but to give you a place to start. You can begin with any of these behaviors; your dog may find one kind of task easier than another at first. You can work on all of them, even in a single session, if you want. Experiment!

If you run into problems, look first at the "Clicker Tips" and "Frequently Asked Questions" sections of this book. It's probable that other people have had the same problems and questions when starting out. Use the resources available. You can teach yourself this art, by trial and error and study. Many people have done so already.

Remember, too, that each training session with each dog is different. You have to "wing it" and use your imagination. Sometimes pets catch on faster than people do! Be patient with yourself, and have fun.

The Basics

Operant conditioning is an active, participatory experience, like dancing, or making music. You won't learn how to do it by reading about it, or by arguing about it. You need to try it. For that you need a pet and something the pet likes to eat and two or three minutes of spare time.

To get ready for a clicker-training session, prepare twenty or thirty cut-up treats, something very delicious: hot dogs, leftover chicken, mild but firm cheese—clicker trainers use all of these. Make the pieces small: I suggest pea-size. If you have a Maltese Terrier, one piece is big news. If you have an adult Great Dane, give several pea-sized bits at once, so the dog knows there's something in its mouth.

Don't use commercial products at first. Later, you can carry the commercial treats, which are clean and dry, in your pockets, and use them, for example, to reinforce previously trained behavior in new surroundings. But right now, while you and your dog are learning new ways to communicate, I'd suggest fresh, real food.

In addition to being only somewhat delicious, most of the commercial products take some time to chew. Is that a problem? Yes, indeed: clicker training is all about TIMING, and that includes the rhythm of the shaping interchange, which is part of the learning. If the rhythm is constantly interrupted (by, for example, irregular pauses for mastication) both the dog AND YOU will soon feel like doing something else. "Let's see what's on TV" (you); "Let's see if any other dogs were in my area last night," (the dog). So: make sure, at least at first, that your payoff is some kind of food that your dog tends to gulp down whole; then both of you can get on with the communication.

The First Session

Your first session may be only a few minutes long, but you need peace and quiet in those minutes. After all, you (not just the dog) are learning something new. You need to be free of meaningless interruptions.

Pick a place where there are no distractions and where you both feel at ease. The living room is fine, or the kitchen. (Outdoors there are too many other things for the dog to think about.) Wait until you and the dog can be alone; the LAST thing you need is kibbitzers telling you you're doing it wrong. If you have other pets, put them elsewhere temporarily.

Put the treats in a dish or bowl where you can reach them but where the dog cannot. Be prepared to move around while you're training; don't always just sit on the couch. If you are moving, your dog will find it easier to get moving, too.

Some dogs are not very interested in food treats at first. That's fine; once they catch on to the meaning of the clicker, a zest for treats will develop naturally. It is always a good idea to have your first brief click sessions right before the dog's usual meal time. Then the dog's normal appetite is working in your favor, to help you gain the dog's interest and participation.

How about Puppies?

You can start clicker-training a new puppy the day you get it home. Unlike conventional training with a collar and leash, there's nothing frightening or potentially harmful in clicker training, so you don't have to wait until your puppy reaches some specific age to begin. (Furthermore puppies, being little food-seeking missiles, LOVE finding out how to make you click, and often learn faster than older dogs.)

You can, of course, start clicker training with a dog six months old, or six years old, or sixteen years old. They can all enjoy this game. Also it doesn't matter what breed of dog you have, or what sex, or what temperament: clicker-training is fine for hyper enthusiastic bouncing-off-the-walls dogs, and for quiet, dignified, or timid souls too.

Here are several different starting exercises for you and your dog to practice on: "Sit and down," "Come when called," "Walk on a loose leash" "Target training," and "101 things to do with a box." You may begin with any one. You may try all of them. Some dogs take to one behavior more easily than another; if, after three or four short sessions, you don't seem to be making progress, feel free to shift to a new behavior, and then come back to the old one later on.

Clicker Training the Sit and Down

To begin: Get your treats ready. Click and simultaneously hold out a treat right under your dog's muzzle, so it can get it easily. Move a little: repeat this several times. These few first clicks and treats are just to help the dog understand what the click means. Vary the length of time between the click and the arrival of the treat: sometimes immediately, sometimes a beat or two later. Always give the click first, and then the treat. Keep your treat hand quiet, or behind your back, until you click, so you don't inadvertently give the dog false cues by unconscious hand movements. Sometimes you might want to wait a few seconds after you give the treat, before you click again, to help the dog become more aware of the importance of the click.

Now click and toss a treat to the floor (or onto a plate) where the dog can see it. Help the dog find the treat if necessary. You are showing the dog that the click always means food is coming, BUT it may not come in the same place every time, and it may not come right away. Do this two or three times.

Some clicker trainers like to condition the dog to the clicker by repeating this introductory process many, many times. I prefer just to get on with it. As soon as the dog is pricking its ears when it hears the click, I think it's time to give the dog (or any animal) the most important information of all: that its own actions can cause the click to happen.

Step One

To start clicking the sit: hold the clicker in your dominant hand, the treats in the other. Holding the treat in your closed hand, move your hand backwards over the dog's head, between its ears, so it looks up. Click the upward look. Repeat. Back up a step or two, coax the dog toward you, and click if it looks up at your face. Repeat.

Now lean over the dog just a little. This should make the dog lean backwards so his tail end lowers. Click that movement. Treat. Repeat. Work fast and click often; it's better to give a click for nothing much (eye contact, say) rather than let time go by with no rewards so the dog loses interest. Watch the hind legs; when they start to fold, click.

At some point the dog will sit. Click and give the treat immediately. Back up a step, coax the dog to you, wait, and see if the dog will sit without any hints from you. Click. Repeat. Move sideways a step or two. Smile at the dog, catch his eye, and pause-the dog will usually try his 'lucky sit' again. Click and treat!

When you click, and make the treat available, the dog will usually jump up again. That's FINE! Contrary to conventional training, we don't care what the dog is doing when we feed it: only when we click. Your click "marks" the behavior you want. The dog will remember what it was doing when you clicked, without any help from you.

By the way, if the dog doesn't sit at first, be patient. Use your closed hand and the treat to try luring the dog's head backwards again. Don't push the dog. Don't tell it what to do. ("Don't tell it what to do?! How will it know what to do?!") Don't worry. In clicker training, we don't start by ordering the dog around. First, we get the behavior. Later, we'll name the behavior.

What if the dog jumps all over you? Ignore that behavior. Wait for it to die down. When the dog puts its front feet back on the ground, start the hand movement again.

Step Two

Now we're going to increase the length of time the dog will sit. Wait until it sits. Click. Treat. The next time the dog sits, don't click right away. Move your treat hand away. Is the dog still sitting? Click. Treat.

By delaying the click a little you are teaching the dog to sit there and stay there until it hears the click. You do not need to keep clicking to encourage the dog: for now, the click ends the behavior. By withholding the click you can extend the duration of the behavior. Don't talk to the dog; don't gush over it; that will just add confusing non-information. Be quiet.

Ending the Session

Research has indicated that many short training sessions are much more effective than a few long ones. Your first sessions might be no longer than four or five minutes. Quit while you and the dog are still having fun. You may get no further than Step One or Step Two in the first lesson, or even the first two or three lessons. That's fine. On the other hand, you may find yourself with one of those whirlwinds that wants to play this new game for half an hour or more, or that goes all the way through Step Six in the first lesson. That's fine too. Some dogs might give you Step Three or Four before you have accomplished Step Two. That's fine; go ahead with the step you've got, and go back to the earlier step later. (If you only got in two clicks for some desirable behavior, that's fine! Each click counts). The important thing is to make progress, however small.

If the dog quits on you (as they may, at first, when they are just learning this game) or if YOU get confused or feel yourself getting impatient or angry, stop at once and try again later. You'll find that an amazing amount of learning goes on between training sessions, especially if you keep them short. You may find that a problem that eludes you today will straighten itself out tomorrow.

Step Three

When the dog is initiating the "sit" on its own, when the dog can sit for a few seconds without your hand in place, you can start using your treat hand in a new way: to tempt the dog into lying down. Put your treat hand, fist closed, treat inside, to the dog's nose, as it is sitting. Then move it down, slowly, between the dog's front paws, very close to the dog. The nose will follow the hand, the dog is likely to lean backwards, and then to start to lie down. Click when the front legs start to buckle, the first time. Treat. The second time, lure it a little farther down, pulling your hand away gently when it touches the floor. Does the dog lie down? Good.

Click. Does the dog pop up, then? Fine. Doesn't matter. Give it the treat while it is standing. Then lure it down again.

Step Four

At some point—and this is your judgment call—gently reduce your luring motion downward. Let the dog know you have food, but don't lead its nose down. Wait. You are waiting for the dog to try lying down, ON its OWN.

You don't need to wait for the complete "lying down" behavior. Click if its paws just start to slide out, or its front elbows start to bend; or even if the dog dips its head toward the floor. Your click (followed by a treat) communicates to the dog that it is thinking along the right lines: going floorwards. Wait again; give it time. This time, let it go a bit further. Maybe all the way down.

Repeat. Watch the dog. You may actually see, on the dog's face, the moment when "the lightbulb goes on;" the moment when it realizes that if it lies down, you will click! Even tiny puppies six or eight weeks old can figure this out. That's when your pup will begin to lie down on purpose. The dog may even fling itself to the floor. If that happens, you can signal your pleasure not by clicking repeatedly— always give one click only—but by giving it a whole handful of treats: a dramatic event we call a jackpot.

As trainer Lana Mitchell puts it, for you this is a training game; but it's a thinking game for the dog. When your dog realizes, consciously, that by lying down, and waiting, it can make you click—when the dog discovers it can EARN its own treats—it has learned something wonderful that we can build on in many, many ways. It is an important first lesson; and an exciting one for the dog.

Training Note: Start a New Session with a Review.

It's always a good idea to start each session by going back a step: If you got to Step Five, and the dog was lying down on its own, begin the next session with a quick review of Step Four, luring the dog into a down, then waiting and clicking for a partial down, then for a full down on its own. Two or three "review" clicks will usually bring the dog up to speed so you can go on.

The next steps for any behavior, including "Down," are to extend the duration of the behavior, to add distractions, and to attach a cue: a hand signal, or a word, or both. To proceed with these steps, turn to page 30, "Increasing your Clicker Training Skills."

Clicker Training "Come When Called"

Coming when called! The bugaboo of all pet owners. Training a reliable "Come" is simple but it's not easy. It takes thought, attention, and some common sense. The clicker will be a great help.

Step One

Sit on the floor. Coax the dog over to you. When it gets to you, click and treat. Get up. Move. Try again. Repeat.

Call the dog when it is in another room. Click and treat when it gets to you. Do this at odd times of the day and when unexpected. ALWAYS click and treat when the dog responds. If the dog does not respond, don't give it a second chance. Let it see you put the clicker and treats away (or, even better, let the dog see you click and treat another dog, or the cat.)

Children can help; put the children on both sides of the room and have them do this: One child speaks the dog's name. When the dog looks, the child says "Come," and pats the floor or whatever is necessary to get the dog to come. (Don't wave treats! Hide the treats.) Click when the dog arrives. Then the child gives a little treat and the other child calls the dog.

Even a puppy soon thinks this is a great game, romping back and forth between two kids with clickers and treats.

Step Two

Try it outdoors. If the dog is a real bolter, use three people. One holds the dog on a long leash, and the other two call the dog back and forth between them. Remember that the important element for developing reliability is not the food, not the call, but the click when the dog gets to the person. That is when the learning occurs.

Take the dog for walks. A "Flexi" leash, which winds itself up into a little box, is nice for this exercise, because it allows the dog to get quite far away from you, feeling quite free, without actually being free. Every now and then, call the dog, from near or from far, and click and treat when it comes to you. If it ignores you, walk on another ten or twenty paces and try again.

If the dog pulls constantly, shorten the leash, stand still, and let the dog pull until it gives up. Then call it, even if it's right next to you; when the head turns, click, and treat. Don't yank the dog, or scold it for not coming; that just teaches it that "Come" means "trouble."

Every time you call your dog, and it comes, be sure to click. If you have no clicker with you, make a clicking sound with your tongue and mouth, or say the word "Good" (not as clear, but better than nothing.) Don't let the dog run around loose in public places until you have an enthusiastic return to you well established. This may take weeks, especially if the dog has been ignoring "Come" for years. With a new puppy, it might not take more than a few days.

Step Three: Going Free, Outdoors

When you do let your dog loose, away from home, do it, at first, in a fenced place, and preferably a really boring place. Let the dog have a good sniff around, and then use your "Come" and make it stronger by bending down invitingly. Click, jackpot, and go home. Make that first, one-trial lesson memorable.

If you are outdoors, and the dog doesn't come, go and get the dog. Put the leash on, and take the dog home or inside. Don't give it four or five chances to ignore your call, before it finally comes; you will just teach it to ignore your first four or five calls. Don't punish it when you get hold of it; you will teach it to be hard to catch. Just quietly bring it home. You can, in fact, click and treat the dog for letting you take hold of the collar; this defuses any tendency to stay out of reach.

When you are up against extremely powerful distractions, such as a park full of squirrels, you can use the distractions themselves as a reinforcer. For example: take the dog to the park on a long leash. Let it sniff around. By and by, call it: if it comes, click and then, instead of giving a treat, unsnap the leash and let the dog chase squirrels. Help it, even!

Then go catch the dog, put the leash on, and go home. It seems counter-intuitive, I know, but each experience of this reinforcer will actually strengthen the behavior of coming when called.

Sometimes, with a dog that has begun coming regularly and well, you arrive at a day when the dog suddenly seems totally deaf to your voice. You have reinforced coming when called many times, and now, suddenly, it doesn't seem to work. There is a negative move that you can make: Call the dog; give it an opportunity to succeed; and when it does not give the behavior, take the dog straight home, without pausing for any more sniffs or fun. The next time you go out, you may see a return to its previous responsiveness. If not, go back a few steps, review the process, and make the task easier, with fewer distractions, and more opportunities for clicks and treats.

Step Four: Maintaining "Come"

"Come" is such an important behavior that I think it warrants attention throughout a dog's life. Use daily events to reinforce the "recall," the dog's quick return to you. For example: now and then, when you are going to give the dog its dinner, say, or a new toy or a chew bone, or when you are going to play with it, call the dog from wherever it is. When it comes, click, and then give it the dinner or the bone, or toss the ball or the Frisbee.

Most dogs love to ride in the car. If you are planning to take the dog somewhere in the car, use "Come," and reinforce the behavior by opening the car door and letting the dog get in. As long as you own the dog, make "Come" very good news by following it, every once in a while, with a happy surprise. Someday this practice may stand you in very good stead.

Walk on a Loose Leash

The technique we are engaging in is called "shaping." Instead of striving for one complete behavior from the beginning, as we do in conventional training, we "build" the behavior, bit by bit. There are many different ways to shape any given behavior. Here are a number of ways to teach a dog to walk on a loose leash. If you have a confirmed puller already, you have some un-training to do, which will take a little longer. Experiment with these suggestions: train with your brain, not with your muscles.

Step One

Take the puppy or dog to a quiet room or hall where you have enough room to walk around. Coax him to your left side, by patting your thigh and speaking his name. When he comes toward you, click. Give a treat.

Step forward with your left leg; coax the dog with a word or by patting your leg. DON'T lure him with the treat; he will only learn to follow treats. You want him to learn to follow YOU. If he comes along with you, click and treat. Stop to treat, don't treat him "on the move;" it confuses the issue to have eating occurring simultaneously with the behavior the dog is supposed to be learning. Where the dog is when he gets his treat should be irrelevant. What matters is that he is more or less beside your leg when you click.

Start up again; if the dog starts right up with you, click at once, stop and treat. Now take three steps before you click, and click if the dog is next to you. Turn around, coax the dog to come with you, and go back the other way. Click every three steps. If the dog is right with you, start clicking even four or five steps.

Step Two

Add variations. Go for longer numbers of steps. Try going a little faster, or a little slower. Make it easy for the dog to "win;" don't scold him if he moves away, just click him when he comes back.

Click the dog for staying close to your leg when you are turning, or reversing direction. Try stopping; click the dog for staying close when you stop. Go faster and slower. Keep those clicks coming every five to ten seconds, but be sure you have a reason for each click. Keep your sessions short, a few minutes is plenty.

Step Three

Move outdoors. For safety's sake, put the dog on a leash; but tie the leash to your waist so you won't be tempted to yank the dog around, or to pull back if he pulls. (Most real plungers have actually been trained to pull forward by their owners constantly pulling back while nevertheless letting the dog go wherever it wants.) Now go back to coaxing, and clicking the dog every three steps, then every five, or ten, or whatever seems reasonable. Use your judgment; count how long the dog is staying with you as a rule, and make that your base line.

When the dog can go a long way, straight, without losing interest in you, start adding corners and curves and variations. Dogs love this game, and will begin to try to outguess you by staying close to your leg no matter WHAT you think up to do. Make the game just hard enough to allow the dog lots of success.

What if the dog sees or smells something he wants to investigate, charges to the end of the leash, and starts pulling? Stop. Stand still. While the dog is pulling, nothing happens. When the dog slackens, click. Treat. Over many walks you can increase the level of temptation. Let the dog discover that when he pulls, nothing happens; but if the leash is slack, you will move toward the other dog, the new person, the hydrant, or whatever he wants to investigate. Your aim, with the clicker, is to give the DOG the responsibility for keeping the leash slack. By and by it will become a habit, and you will no longer have to reinforce with praise or clicks more than once or twice on a walk. And, eventually, you can expect good "leash manners" as a matter of course.

A Step for Confirmed Pullers

The simple process of clicking every few steps, and stopping to treat, repeated on a few different walks, is often enough to convert inattentive pullers into mannerly walking companions. Be sure to stop and let the dog sniff and investigate, from time to time. That pleasure should be permitted to the dog as a reward for not pulling, instead of "stolen" by the dog's dragging you to each thing he wants to smell.

What if the dog hits the trail pulling, and ignores you, in the great outdoors? Tie the leash to your waist. If you have a big dog, wear sturdy shoes so he can't pull you off balance. Now you can use his desire to explore, as a reinforcer. He pulls. You stop, plant your feet, and go nowhere. Say nothing. When the dog looks at you, pat your leg. If he slackens the leash or comes to you, click and move forward. Immediately he'll probably plunge to the end of the leash again, especially if he's been doing that for years anyway. Repeat. Stop: slack leash, click and go.

You may not be able to go more than one step at a time, in the first session. That gets old in a hurry. Be prepared to quit early. When you get tired of this step-and-stop business, put the dog back indoors, or in the car. Try again the next day. Sooner or later the dog will capitulate, and start waiting for you instead of plunging ahead. THEN you have something to click and treat.

An Easy Step

Sometimes you can cure a puller easily with one simple rule for yourself: Every time the leash goes slack, I will click and treat. Take the dog for a walk on the leash, in your usual fashion, but, again, tie the leash to your belt. When the dog in its investigations happens to let the leash go slack, click and treat. By the end of a ten-minute walk, you may have a reformed character.

Guess what? You can do this with two dogs. The rule is, "When BOTH leashes are slack, I click and treat." Pay sharp attention; you may have a habit of letting the dogs pull, that is so ingrained it's hard for you to notice those brief moments when they are not pulling. Did they stop to smell a bush, or scratch, or relieve themselves? It doesn't matter why. If one dog stopped, and then the other, click. Treat them both. Again, this can be an almost instant fix. Two or three five-block walks, over two or three days, might be all that's needed, if your timing is right.

The Gentle Leader

The Gentle Leader is a sort of halter for dogs, like a halter for horses. It goes around the nose and neck. The leash fastens under the jaw. The dog can still open its mouth comfortably;

this is not a muzzle. However, if the dog pulls, its head is turned back or down so it can't see where it is going. So it stops pulling. You don't need to do any training at all. Just hold the leash and walk along. The dog may fight the collar at first, just as puppies fight a regular collar and leash at first; but it will soon teach itself to keep the leash loose. This is an excellent temporary fix while you are retraining the dog; or it can be a permanent help to a frail owner or a child who must walk a big dog. Gentle Leaders are available from pet stores and various Web sites.

Target Training

One of the most useful skills your clicker dog can learn is to touch a target, such as the end of a stick. One of many uses of targeting is that you will be able to move the dog anywhere you like—off the couch, into the car, onto the grooming table, out from under the bed—without having to push, pull, or shove it, and with the dog's eager and interested cooperation.

Step One

Commercial folding aluminum target sticks are available on clicker training Web sites, but you can use anything: a dowel, a yard stick, or a two to three foot long switch from an apple tree. For a tiny toy dog (or a cat) you might use a pencil or a chopstick.

Get your stick, treats, clicker, and dog. Rub a treat on the end of the stick and hold it out beside the dog's nose. When the dog sniffs the stick, click, take the stick away from the nose, and give a treat. Repeat. Move the stick up an inch, down an inch, to the other side. Click for touching or even looking at the stick.

After a few clicks for touching, stand up and coax the dog to walk with you, holding the stick in front of him so he can touch it easily while he's walking. Somehow dogs seem to catch on to targeting more quickly when they are moving. Click for each touch, stop, treat, and continue. Lead it in left and right turns. See if you can get the dog to move around you in a quarter circle, then half a circle, then a full circle.

Step Two

When your dog is enthusiastic about his target stick, start using it for a few minutes now and then, during other activities, to lead the dog around obstacles (a wastebasket, say) or under tables, through table legs, or over small jumps (a broomstick; or your extended leg.) Click and treat generously for the first successes; this is a great confidence builder for timid dogs.

Take the dog outside and try targeting him over and around obstacles in this new and distracting environment. If he loses focus, go back to clicking and treating for just touching the stick from one inch away, or following it for just one or two steps of his paws. Use the target stick to get the dog into the car and out again, or into and out of a box or a wheelbarrow. Use your imagination; every new success is good for the dog and builds his trust in you (and every new success is a reward for you, too!)

Step Three

Try using other objects for targets: margarine tub lids are popular, because they are easy for the dog to see. Put the lid on the floor and touch it with your target stick; click the dog for touching both. Or, hold the lid in your hand, click the dog for touching it, and gradually lower it to the floor and remove your hand. Put the lid in different parts of the room and click the dog for going away from you to touch it, wherever it is. You can use both kinds of targets to teach your dog to negotiate agility obstacles. Extend the time the dog will stay on target. Just click him for holding his nose on the target for longer and longer periods, while you groom him, handle his feet, or clip his nails. The target can be even be used at the veterinarian's office to help the dog stay calm and hold still during medical examinations and treatment.

101 Things to do with a Box: A Good Exercise for an Older, Suspicious, or Previously Trained Dog

This training game is derived from a dolphin research project in which I and others participated, "The creative porpoise: training for novel behavior," published in the Journal of Experimental Analysis of Behavior in 1969. It has become a favorite with dog trainers. It's especially good for 'crossover' dogs with a long history of correction-based training, since it encourages mental and physical flexibility and gives the dog courage to try something on its own.

Step One

Take an ordinary cardboard box, any size, cut the sides down to about three inches, and put the box on the floor. Click the dog for looking at the box. Treat. If the dog goes near or past the box, even by accident, click. Next, after you click, toss the treat near or in the box. If the dog steps toward the box to get the treat, click the step and toss another treat. If he steps into the box, great, click again, even if he is eating his previous treats, and offer him another treat in your hand.

Sometimes you can cook up a lot of "box action" in a hurry, this way: Click for stepping toward or in the box. Alternately toss the treat in the box and hold the treat out in your hand so the dog has to come back to you. If the dog is reluctant to step in the box, and so doesn't eat that treat, it doesn't matter: he knows he got it. If treats accumulate in the box, fine. When he does step into the box, he'll get a jackpot. If you decide to stop the session before that happens, fine. Pick up the treats in the box, and put them away for a later session. Remember, never treat without clicking first, and always click for a reason: for some action of the dog's.

If you need more behavior to click, you can move yourself to different parts of the room so the box is between you and the dog, increasing the likelihood of steps in the direction of the box. Don't call the dog, don't pat the box, don't chat, don't encourage the dog, don't "help" him. All of that stuff may just make him more suspicious. Click foot movements toward the box, never mind from how far away, and then treat. If you get in five or six good clicks, for moving in the direction or near or past the box, and then the dog "loses interest" and goes away, fine. You can always play "box" again later. In between sessions, the reinforcements you did get in will do their work for you; each little session will make things livelier the next time.

You are, after all, teaching your dog new rules to a new game. If you have already trained your dog by conventional methods, the dog may be respecting the general rule, "Wait to be told what to do." So the first rule of this new game, "Do something on your own, and I will click," is a toughie. In that case, the box game is especially valuable, and the first tiny steps are especially exciting—although they would be invisible to an onlooker, and may right now seem invisible to you.

End the first session with a "click for nothing" and a jackpot consisting of either a handful of treats, or a free grab at the whole bowl. Hmm. That'll get him thinking. The next time that cardboard box comes out, he will be alert to new possibilities. Clicks. Treats. Jackpots. "That cardboard box makes my Person behave strangely, but on the whole, I like this new strangeness. Box? Something I can do, myself? With that box?" Those are new ideas, but they will come.

If your dog is very suspicious, you may need to do the first exercise over again once, or twice, or several times, until he "believes" something a human might phrase thus: "All that is going on here is that the click sound means my Person gives me delicious food. And the box is not a trap, the box is a signal that click and treat time is here if I can just find out how to make my Person click."

Step Two

Whether these things occur in the same session or several sessions later, here are some behaviors to click. Click the dog for stepping in the box; for pushing the box, pawing the box, mouthing the box, smelling the box, dragging the box, picking up the box, thumping the box—in short, for anything the dog does with the box.

Remember to click WHILE the behavior is going on, not after the dog stops. As soon as you click the dog will stop, of course, to get his treat. But because the click marked the behavior, the dog will do that behavior again, or some version of it, to try to get you to click again; so you do not lose the behavior by interrupting it with a click.

You may end up in a wild flurry of box-related behavior. GREAT! Your dog is already learning to problem-solve in a creative way. If you get swamped, and can't decide which thing to click, just jackpot and end the session. Now YOU have something to think about between sessions.

You may on the other hand get a more methodical, slow, careful testing by the dog: the dog carefully repeats just what was clicked before. One paw in the box, say. Fine: but right away YOU need to become flexible about what you click, or you will end up as a matched pair of behavioral bookends. Paw, click. Paw, click. Paw, click. That is not the way to win this game.

So, when the dog begins to offer the behavior the same way, repeatedly, withhold your click. He puts the paw out, you wait. Your behavior has changed; the dog's behavior will change too. The dog might keep the paw there longer; fine, that's something new to click. He might pull it out; you could click that, once or twice. He might put the other paw in, too...fine, click that. Now he may try something new.

And? Where do we go from here? Well, once your dog has discovered that messing around with the box is apparently the point of this game, you will have enough behavior to select from, so that you can now begin to click only for certain behaviors, behaviors which aim towards a plan. It's as if you had a whole box of Scrabble letters, and you are going to start selecting letters that spell a word. This process is part of "shaping."

Step Three
Variations and final products: What could you shape, from cardboard box behaviors?

Get in the box and stay there. Initial behavior: Dog puts paw in box. Click, toss treats. Then don't click, just wait and see. Maybe you'll get two paws in box. Click. Now get four paws in box. Get dog in box. Options: Sitting or lying in box; staying in box until clicked; staying in box until called, then clicked for coming.

Uses

Put the dog to bed. Put the dog in its crate. Let children amuse themselves and make friends with the dog by clicking the dog for hopping into a box and out again (works with cats, too.) One third grade teacher takes her Papillon to school on special events days, in a picnic basket. When the basket is opened, the dog hops out, plays with the children, and then hops back in again.

Behavior: Carry the Box

Initial behavior: dog grabs the edge of the box in its teeth and lifts it off the floor.

Uses

Millions. Carry a box. Carry a basket. Put things away: magazines back on the pile. Toys in the toy box. A dog that has learned the generalized or generic rule, "Lifting things in my mouth is reinforceable," can learn many additional skills.

Behavior: Tip the Box over onto Yourself

I don't know what good this is, but it's not hard to get; it crops up often in the "101 things to do with a box" game. If the dog paws the near edge of the box hard enough, it will flip.

My Border Terrier, Skookum, discovered that he could tip the livingroom wastebasket (wicker, bowl-shaped, empty) over on himself, so that he was hidden inside it.

Then he skooted around in there, making the wastebasket move mysteriously across the floor. It was without a doubt the funniest thing any of our dinner guests had ever seen a dog do. Since terriers love being laughed with (but never At) clicks and treats were not necessary to maintain the behavior once he had discovered it; and he learned to wait until he was invited to do it, usually when we had company.

Increasing your Clicker Training Skills

Adding variations

To keep the training interesting, and to make progress instead of getting stuck where you are, you need to keep adding to your requirements for clicking: you need to make the behavior more complex, by asking the dog for more, before you click. For example, here are some ways you can increase the complexity of the simple "down."

Move away from the dog, and click it for staying down even though you have stepped back.

Move left, right, center; jump up and down; do something noisy or silly; click the dog if it stays down anyway. Walk around the dog; touch the dog and then step back again. Click it for staying "down." If it gets up, wait for it to "down" again. Then click for going down (and treat, of course). Then try your maneuver again, and give it another chance to experience staying down during your movements.

Play the "down" game in new places, such as outdoors. Don't be afraid to lure the dog down, once or twice. Then wait for it to go down, by itself, and click that. The first time you try clicking for "down" in new territory, you need to go "back to kindergarten" and make the task very easy for the dog.

Extend the length of time the dog has to lie down. Keep it varied: sometimes five seconds, sometimes twenty. Don't ask for more than thirty seconds from a little puppy. For a three-minute or five-minute down with no reinforcement, the dog needs to be old enough to tolerate boredom.

Establishing a Cue

Sometimes a dog who has learned one behavior, such as "down," will start "throwing the behavior at you" all day long: rushing up and crashing at your feet wherever you may be, in the hopes of getting a treat. This is the perfect time to teach the dog a signal, or cue: a name for the behavior, such as the word, "Down."

You are going to show the dog, over a period of days, that lying down is still good for a click, but only after you have spoken the key word (or given the key signal) that means "down." You want the dog to discover, through repeated experiences, that going down on its own no longer leads to clicks and treats; but that it can be sure of getting a treat IF it waits for the right cue before giving you a "down."

Say "Down," just before it is going to lie down anyway; click and treat. Coax it to get up. Repeat. After cuing it and reinforcing the down after the cue, two or three times, wait. Let it lie down. Do nothing. Coax it to get up. Say, "Down." Click the next down. Repeat this process, clicking for downs after the word, and ignoring downs that were not cued with the word. Repeat daily, eight or ten times, for a week. This is "respondent" conditioning. At first it goes more slowly and takes more repetition than operant conditioning. Be patient; once the dog has learned two or three cues by repetition, with the cue gradually sinking in, the dog will be able to generalize: to understand the nature of a cue. Then subsequent cues will be learned more rapidly.

If you want to use a hand signal (many dogs respond better to hand signals than to words, at first) just replace the spoken cue in the instructions above with your chosen hand signal. A hand flung straight up in the air is the standard obedience cue for "Down."

When the dog is responding well to the signal, try signaling from further away, or when the dog is busy with other things, or in distracting circumstances. Your all-powerful clicker will be a great help in explaining to the dog that there is value in watching out for those cues or signals and responding to them any time, anywhere. (For more information about cuing behavior, see Chapter 3, Stimulus Control, in my book, *Don't Shoot the Dog*.)

Training Multiple Behaviors

You don't need to limit your clicker training to the program outlined above. Right from the beginning, you can also leave clickers lying around the house and click the dog for other good behavior, whenever it's convenient. For example, it's very helpful to click a puppy for relieving itself in the right place. Housebreaking becomes infinitely easier with a clicker and treats by the door or in your coat pockets.

You can also click your puppy or older dog for other behavior, such as standing still while you groom it. You will not confuse the dog by clicking for several behaviors. The clicker means "You win!" The dog will be delighted to learn that there are MANY ways to get its person to click. That will make the dog feel very smart!

Find a Training Partner

One good way to explore clicker training is to find a friend who also has a pet, and get started together. This chapter has given you several behaviors to try. If you and your friend each pick a different starting point, you will make discoveries you can share with each other; and you can reinforce each other for successes.

Team training is also very useful. Sometimes an observer can see what the dog is doing more easily than the handler, for example when training a dog to trot straight for the show ring, or to sit squarely for obedience work. One person can handle the dog and the treats, and ask for the behavior. The other person watches, and clicks when the dog moves correctly. Then the handler treats.

You can take turns being the handler and the observer. You can handle your own dogs, or each other's. This way, the observer can concentrate entirely on studying the dog, and on clicking with perfect timing, while the handler copes only with leash, food, and dog. Team training speeds up YOUR learning, and gives the dog more accurate information. Besides, it's fun.

Advance your Skills: Use the Resources Available.

There is much more to operant conditioning than the basic shaping techniques outlined here. If you are developing a working dog, or a competition obedience dog, you should know about behavior chaining, instructional reinforcers, stimulus transfer, and many other marvelous clicker training tools. See *Resources* for suggestions about books and videos that you may find interesting as you gain more skills.

Clicker Tips

If you are working with the "Getting Started" section, these tips will serve as a review. They cover many of the common mistakes beginning clicker trainers make.

Always click first, then treat.

In this kind of training, you do not need to keep the food in front of the dog, to bait or lure the dog. What your pet is doing WHEN THE TREAT ARRIVES is not important. Your pet will remember what it was doing WHEN IT HEARD the click. After you click, you can let the behavior stop, give the treat, and then go back to the training, without causing problems.

Always click while the behavior is happening.

The "click" is like an acoustic arrow; it goes right into the animal's nervous system with the message, "What your body is doing at this instant has just paid off." Use it at the moment that counts. If you are teaching your dog to jump, for example, don't click before the dog goes over the jump. Don't click after the dog has completed the jump. Don't click when you think it's going to jump. Click when the dog is in the air. Once the dog is jumping, you can improve the way the dog lands, for example, by clicking at the moment of landing correctly. The food can come later; the click carries the message.

Only click once.

Resist the temptation to make multiple clicks—clickety clickety clickety—for specially good behavior. The timing of the click is crucial; it is when you click that tells the animal exactly what you liked. If you click over and over, the animal can't tell which click is the meaningful one, and the behavior will not improve. Furthermore, your pet might decide only multiple clicks are worth working for.

Don't use your clicker to get the dog's attention, or to get the dog to come.

Sure, it will work that way, at first, anyhow. If you've had trouble with those things in the past, that new response may be reinforcing for you, so you're tempted to try it over and over. Watch out, though!

If you continue to call your dog with the clicker, or to click to try to distract the dog from other interesting things, you will actually be reinforcing the behavior of staying away, or of looking for distractions. The problems will increase.

Don't use your clicker for encouragement, or as a start signal.

You should always be thinking, "What am I actually reinforcing?" If your dog hesitates, or lags behind you, and you click to encourage it, you are reinforcing hesitation or lagging. Click only for behavior you want.

Use the right size and type of food treat for your pet.

Here are some common-sense rules about food treats:

Start your training when the pet is hungry. Many dog owners do a five-minute clicker training session before the dog's breakfast or dinner.

Don't leave food down all day long; feed your pet at specific times. If you need to leave food available most of the time (as vets recommend for most birds and small animals) take the food away for two or three hours just before your training session.

Make your treats part of the pet's total food intake; if you are using a lot of treats, cut back on the pet's normal food a little.

At first, while you and your pet are both learning this new game, use a highly preferred, delicious-smelling food, not boring old kibble or pellets.

Keep the training fun. Always toss in a few easy tasks when you are working on hard ones.

Vary the Difficulty of the Task

I hated my piano lessons, as a child. Whenever I mastered a piece, instead of getting to have fun playing it, I just got a newer harder piece to struggle over.

Don't make your sessions go from old easy stuff to new hard stuff every time—that's a great way to teach your pet to hate training and want to quit early. Particularly when you are extending the duration of a behavior, go back and forth between easier and harder. Are you teaching the dog to stay on its mat for 30 seconds? Click for ten seconds, then 25, 15, 35, 10, 40, and so on. Keep the dog hopeful of success.

Always end the session with something easy and fun so that the animal can be sure of earning reinforcement.

Keep your training sessions short and varied; don't practice the same thing over and over.

Train in short bursts. Three sessions, each five minutes long, will do you more good than one half-hour session in which you and the pet both get tired. You do not need to drill or practice this kind of training to make it "sink in." Experienced correction-style trainers are often amazed to find that behavior trained by shaping and reinforcement does not deteriorate. With your clicker you can communicate to your pet. The pet will remember whatever you have communicated successfully, and may eagerly offer to do it again, even if days, weeks, or months have passed between sessions.

Expect to progress in small steps.

In generic or traditional training, we ask for the whole behavior from the first ("Heel!" meaning "Walk right next to me wherever I go") and correct any departure from that behavior, until the animal can do the behavior for prolonged periods without "needing correction." Shaping behavior with reinforcement is different. We build behavior in little steps. The animal discovers how to earn reinforcers, at each new step. It takes more brains and thought from the trainer but the rate of progress can be spectacular.

Nevertheless, sometimes the learning "hits a plateau," where the animal doesn't seem to make any improvement for quite a while. Hang in there: plateaus often precede a huge leap forward. It helps to start each training session with a little review.

If a behavior your pet knows well suddenly falls apart, "Go back to kindergarten."

Sometimes a change in circumstances makes more difference to an animal than we might think. For example a dog that heels beautifully, or comes when called perfectly, in your house or yard or at the training school, may act as if it never heard those words before, in a new place. Don't scold! Instead, pretend you never trained the behavior, and go back to the first thing you ever clicked for, even if it was "Come from three feet away, for a click and a piece of roast chicken." Then proceed up through the increasing difficulty of the task, just the way you first shaped it, but in the new circumstances. (You can keep a leash on the dog while you do this, for safety's sake.)

We call this "Going back to kindergarten." This quick review of the shaping procedure is sometimes all you need to establish the behavior in a new environment. It may take one session, or several, but it will certainly go much faster than it did the first time. Be sure to give a jackpot of great treats if it works in one training session!

Don't think you can read the dog's mind.

His feelings, yes: we can sometimes tell from a dog's attitude how he feels about things (sometimes we're wrong, too.) But why he feels some way, or why he is doing something, is guesswork at best.

When the dog does not do what you expected, you can always find people who will insist it happened BECAUSE of some reason. The dog is "stressed." The dog is "ring-wise." The dog is "trying to get back at you." The dog is "too dominant."

The dog "must have been abused," etc. Then they tell you to how to FIX their imagined cause, often by punishment.

Don't listen! If a behavior your pet knows well suddenly falls apart, don't spend a lot of time wondering why. You may never know the reason. Maybe the dog thought the rules only worked if you were wearing white shoes, and you are now wearing your new purple Adidas. Just go back to kindergarten, review the shaping, and reinforce easier steps, until the behavior reappears.

Quit while you're ahead.

It's always tempting, when we get a wonderful behavior, to try to make it happen again and again, until the subject makes a mistake and the behavior breaks down. Learn to quit when you get something good. Leave a success in the pet's mind. If you are not making progress, and you want to stop training, go to something easy that you know you can click the dog for before you stop training.

Have fun!

If you find yourself getting frustrated or angry, stop right there. You cannot do this kind of training when you are angry. Wait and try again later. Trainer Catherine Crawmer says the most important tool in reinforcement training is a cup of tea. When you feel baffled, sit down and have some tea and think. Don't display your anger to the dog: it teaches the dog nothing except that you are unpredictable.

CHAPTER 4

But What About...?
Frequently Asked Questions

"Help! I don't even know where to begin."

Don't begin by trying to fix some problem. The clicker is for starting behavior, not for stopping it. One easy way to start is to "go fishing" for a new behavior. If your dog already knows the behaviors described in Chapter 2, pick some simple trick, like turning in a circle; bowing; shaking hands; or touching a target; and use that as your first practice behavior.

"My dog is afraid of the clicker. I clicked at him and he ran and hid under the bed."

Usually a dog is afraid of the clicker because he does not know what it means. He lacks the correct information. That may make him suspicious (some dogs, not unwisely, associate any new experience with going to the veterinarian.)

Don't worry about the fear; work on giving him the information. You can soften the sound a bit by clicking the clicker in your pocket, or behind your back. You can really muffle the clicker by putting several layers of adhesive tape over the "dimple" in the metal part. Now click—just once—as you put his dinner down; or just as you start to toss him a special treat or give him a special toy. Click (just once, in your pocket) when you let the dog out (or back in), or start out the door on your walk together. Find an excuse to use the clicker to say "Something nice is coming" at several intervals during the day and evening. How many click-events do you think it will take, across three days, for your dog to decide a click is pretty good news?

"What if my dog doesn't like treats, and won't work for food?"

At a trainers' picnic a dog-owner told me his German shepherd was a picky eater and wouldn't work for food—while behind his back the dog was stealing fried chicken scraps from discarded paper plates. I cut up some chicken and clicker-trained the dog to follow a target stick in about five minutes. Moral: to start with, use a preferred food.

It's true that some dogs are legitimately suspicious, or even cynical, about treats: free food in the past may have had bad consequences from the dog's point of view. So, if you have a dog like that, go slow. Just click and treat, click and treat, a few times, twice a day, for no particular behavior, at no particular time.

You could stand by the dog's bowl and toss the treats in, after each click. Even if the dog is very suspicious, it will hear the food fall in. Once the dog decides you have no nefarious plans up your sleeve, it will eat the food. Once it learns a few ways to make you click and treat, the dog will start to love the treat more! Try it and see.

"What if I have more than one dog?"

Separate the dogs—work with one at a time and put the others outside, or in the laundry room, or in their crates, or in the kitchen, or tie them to a chair. Sure, the others will hear the clicks; they will not get the treats, so they will not be confused—just eager for their turn.

"How long should a training session be? How often should I train?"

Five-minute sessions, whenever you can fit them in, are the best way to start. Scientists have proven that dogs, pets, and people learn more in several short sessions than they do in one long session. Short sessions are more fun and less tiring, and easier to fit into a busy schedule, too.

"Can I train more than one behavior in a single session?"

Yes, of course. You can work on come, and sit, and then heel, and throw in a trick or two, in any one session. Changing around is fun. What you should NOT do, however, is work on more than one detail of any given behavior, in a single reinforcement. If, for instance, you are working on "Trotting with your tail held high," (a nice behavior for the show ring) just click or withhold clicks for that; don't suddenly scold the dog for lagging behind. "Not-lagging" is a separate task. Work on it separately.

"I teach dog classes, with a lot of dogs and people in one room. I can't see how we could use clickers, with ten people clicking all over the place."

Most dogs hear better than we do. The way to use clickers in a class is just to start using them. The dogs will be fine. They know where their owners are, who has their treats, and where the sound is coming from. A dog can figure out "That's MY click" in about three clicks.

"Why can't I use a word instead of a click?"

A verbal "click," a word, is not as unique and distinct as the clicker, so it is not as useful for learning new skills and behavior, but it will serve to keep already learned behavior going. Some people also use a special touch—a light tap on the dog's head or back, for instance—as a substitute for the click: very handy in the show ring.

"What happens if I make a mistake in training? Will I ruin the dog?"

If you make a mistake, laugh and pet your dog. It is easy to click too soon or too late, or for the wrong thing, or to miss a great chance for reinforcement, and we all do it all the time. In the long run, you will get enough clicks in at the right times to communicate to your dog what it is that you want. While an erroneous *punishment* can do all kinds of damage to the learning process, an erroneous reinforcement or two is harmless. Clicker-training is creative, interactive, and cumulative; it is a very forgiving system.

"What do I do if the dog makes a mistake?"

Don't click. That's all. Don't say "No," or jerk the leash, or push the dog in the right direction. Adding punishment, correction, or coercion to this system does not help the pet understand how to earn clicks. It does not make the pet more reliable, despite the widespread conviction of many trainers that force is necessary. It definitely does make the dog less interested in learning.

"But doesn't that mean I should never punish the dog? What if the dog jumps up, or nips me, or takes food off the counter, or runs away?"

Some people think that using positive reinforcement means you never reprimand the dog or control it physically. That's unrealistic. Leashes are a fact of life for dogs. It's necessary to keep a dog on a leash when you are going to strange places, or out in traffic, or amongst strange dogs. And of course your dog needs to understand the meaning of "No." You need to interrupt behavior such as mouthing your hands and clothes, for example, or trying to grab food from the kitchen counter. Remember that timing is just as important in correcting misbehavior as it is in reinforcing good behavior. Your response should occur while the behavior is happening, not afterwards, or just before you think it might happen.

While correction and scolding can stop behavior (while you're around, anyway) they are not efficient ways to teach the dog to do something new. For that, clicks and treats work best. As for stealing food, tipping over the garbage, and so on, it's up to you to monitor the environment and put temptation out of reach.

"Isn't it necessary to make the dog respect me, even fear me, if I want long-term reliability?"

You will find that the more things your dog learns TO DO, the more he will trust and respect you, and the easier it will become to make him understand what he should not do. Furthermore, training with reinforcement creates a dog that is doing things because he WANTS to do them, he understands what the job is, and he is confident of a successful outcome. That's true reliability.

"You don't say much about commands; when do I tell the dog what to do?"

When the dog has learned the behavior. First the dog learns that sitting is good for a click and treat. Then the dog learns that sitting only gets a click and a treat when you say "Sit!" So the word becomes a signal that treats are now available for that particular action.

When we are training with punishment—with leash corrections, for example—it is only fair to warn the dog first. Thus we say "Heel," before we pull the dog along with us, right from the beginning. What this command means from the dog's point of view is "Heel or ELSE you will be jerked." The dog learns to do the right behavior to avoid getting pushed or pulled.

With clicker training there is no point in making meaningless sounds at the dog, until the dog has learned the behavior. The command, in traditional training, is a warning signal. The cue, as the signal is often called in reinforcement training, is a green light for a particular action that has paid off many times in the past.

"What if I give the cue and the dog doesn't do it?"

Then you have not established the cue under that circumstance. Here's a common event: A dog may "come" when called beautifully at home, but in a park with squirrels he's suddenly deaf. It doesn't mean he's disobedient, it just means the trainer has not yet asked for and reinforced "Come" often enough under these new conditions. One might have to train (on leash, at first, and step by step) "Come in strange places," "Come from afar," and "Come, even when there are squirrels." You can get there, faster and more surely, by shaping the behavior step by step, than by getting mad, or punishing failure to respond.

"When can I get rid of the food?"

That's a popular question. The click, and the treat, are for learning a new behavior, or for learning to do it in new or tougher circumstances. Once the behavior is learned, you will not need to treat for that behavior any longer, as a regular thing.

"What if I want to go into obedience training, or some other form of competition, with my dog?"

The clicker is a superb tool for explaining to the obedience dog just what the nature of the task is, from perfect heeling to advanced exercises such as scent discrimination. Brainy trainers everywhere are pioneering in developing ways of shaping traditional exercises with reinforcement instead of with correction. New books and videos are in development. More and more trainers are using reinforcement training to develop everything from household pets to police patrol dogs; but everywhere in dogdom it is still a rather new idea. Jump in! The rest of the pioneers need your input!

"How can you use clicker training for obedience, when everyone knows you can't use a clicker or treats in the obedience ring?"

Once again, the clicker and treats are a tool for learning the behavior. Once you get into the show ring, the learning part should be already accomplished. By now, you have substituted a verbal cue for the click, and petting and praise for the treat; so you don't need them in the show ring.

The dogs that "fall apart" in the ring are often dogs that have been trained by correction. As they spend more and more time in competition, they make more and more mistakes in the ring; not because they "know" they won't get corrected, but because the absence of correction conveys the information that what the dog is doing is what he is supposed to do. Ironically, you are allowed to praise your dog in the obedience ring, but you are not allowed to correct him. So training with reinforcers actually gives you an advantage in communicating with your dog in the ring.

"My dog has already had a lot of obedience training. How do I mix clicker training into what he already knows?"

Morgan Spector, an obedience teacher and competitor who has changed from traditional methods to clicker training, calls this a "crossover dog." Crossover dogs (and crossover trainers) have a lot to unlearn as well as a lot to learn. It can be harder, in some ways, for experienced traditional trainers to learn clicker training than for rank beginners. Don't start by working on behavior your dog already knows. Start with tricks or something new. Then you can start clicking obedience skills, after you and the dog are comfortable with the new system.

"Can you use clicker training to train cats or other animals?"

Yes you can. Dogs and horses work to please us (we like to think) or at least to avoid annoying us. Wild animals, and pets such as cats, rabbits, and birds, don't necessarily value our opinion; and if you get mad at them and try to make them do something, they just panic, struggle and scratch, and try to get away. So we say they are "independent," or "disobedient," or "can't be trained."

All animals, however, are able to learn where to go and what to do to find food. Clicker training relies on that natural skill to communicate new information to the animal. So a cat, for example, can quickly learn, via the clicker, where to go—the piano bench, say—and what to do there—hit the keys with a paw—to make you deliver bits of tuna. Who trained who? The cat thinks it did the training. Do you care?

"Does this system work for kids?"

Yes. Telling children when they are doing what you like them to do, and making that rewarding for them, is a lot more effective than telling them what they are doing wrong and making them fix it. The kids like it better, too.

"How about spouses?"

Of course. Everyone likes positive reinforcement. Best of all, sometimes spouses and kids learn to do it back; so you get positive reinforcers too. Remember: clicker training is not something you do to animals or people. It is something you do with them. Everyone loves it; everyone wins.

The Clicker Revolution

Dogs and people learning together

While the science underlying clicker training has been around since the 1940's, and the principles were in use among marine mammal trainers beginning in about 1960, it wasn't until the 1990's that this new technology really began spreading to pet owners. Perhaps my book, *Don't Shoot the Dog!*, the first popular text on operant conditioning, prepared the ground; but I think the change really began when dog owners, not just dog trainers, got that little clicker in their hands and began discovering what they could do with it.

Instead of noticing all the bad things their dogs were doing, people found themselves watching for something that they liked: something to click. Instead of trying to 'fix' the dog, and stop the problems, owners began looking for good behavior, and rewarding that; and the problems usually went away by themselves. The click not only rewarded the dog, it changed the owner's attitude. People discovered that their dogs were smarter than they thought, and lots more fun.

The attitudes of the dogs changed, too. This person with a clicker, who perhaps used to be largely an impediment to whatever the dog wished to do, now became an exciting and valuable acquaintance. A dog that had discovered ways to make the owner click became much more focused on the owner, much more interested in what the owner wanted. Two beings that often had been at odds with each other now became partners, learning together. Clicker magic! And all it took was a few clicks. No wonder people were converted.

Furthermore, training with a clicker didn't require physical skills or years of practice. It was easy to learn. You could pick up the basics from a book like this one, or a video, or on the Internet. As in learning to use email or do word processing or surf the Web, you could get started without a teacher. You could develop skills in brief sessions around the house. You could easily pass on the basics to other people who asked, "How did you do that? Show me!" Seminars, classes, and teaching programs helped to spread the word and develop the technology. The Internet became a huge source of information and support, through Web sites, chat groups, and above all, email listservs, starting with the original Clicker List.

As people discovered how to clicker train, and helped each other to learn more, clicker clubs and centers sprang up all over the world, from Finland to Tasmania, Singapore to Sweden, Russia to Brazil. Wonderful and innovative techniques and applications blossomed independently, in Germany, in the UK, in Australia, and across the US, as one new clicker trainer after another developed his or her own clicker solutions, and then wrote about them or made a video or started teaching in person or on the Internet. By 2002 at least 300,000 people were clicker training according to Internet data. This movement, while based in science, was created of and by people themselves—and their wonderful pets.

Clicking spreads to horses, cats, birds, and other pets

At Sea Life Park, the oceanarium in Hawaii where I began learning working with this technology in 1963, we dolphin trainers developed a vast variety of behaviors, trained with a marker signal, reinforcers, and verbal or gestural cues, in many animals, including wild free-flying sea birds, seals, dogs, feral Hawaiian pigs, jungle fowl, fish and octopi—and of course our own pets at home. We knew that clicking worked for any living being. But in spite of the increasingly widespread use of clicker training with dogs, it was not immediately obvious to conventional trainers that the clicker could be useful for other animals, too.

Alexandra Kurland, a horse trainer and dressage rider, sparked the first breakthrough with her 1998 book, *Clicker Training for Your Horse*. Soon, thousands of people in the United States, Canada, the United Kingdom, and in Europe, especially in Germany, were experimenting with this remarkable new way to teach horses everything they need to know—without spurs, bits, whips, and force. Newly converted riders and trainers were producing articles, books, videos, and new applications. The horses appreciated it, learned fast, and worked much more willingly. Clicker training became a powerful way to put the finishing touches on a highly trained competition horse; to rehabilitate an aggressive or dangerous horse; to train foals and yearlings; and to make any horse safe and manageable. Thousands of amateurs with no conventional horse training background discovered that they could use clicker training to turn their own horse, perhaps once resistant and troublesome, into a cooperative, intelligent partner and friend.

Of course clicker training works for other kinds of pets and domestic animals, too. Clicker training owners have taught dog agility routines to goats and donkeys. Clicker training is the standard method for owners and breeders of llamas and alpacas, popular in the western United States as pets and pack animals and for their wool. Books, videos, Web sites, and listservs are devoted to clicker training for cats. Clicker training for cute behaviors is a delightful way to interact, amuse and exercise an indoor cat, and to replace undesirable behavior with entertaining and healthy activity.

Birds, especially parrots and their relatives, have become enormously popular pets as breeding in captivity makes them more available. The parrot family can be troublesome to manage, though, given to screaming, biting, and pulling out their feathers; and punishment just makes things much worse. Clicker training to the rescue. Using positive reinforcement with a marker signal, even a beginner can soon get on better terms with these bright and interesting but highly emotional pets. Bird clicker trainers have also established Web sites and listservs to help each other out.

The smaller mammals, such as mice, gerbils, rats, and rabbits, make great clicker pals, especially for children of nine years or older. Clicker training a small pet makes a fine science project for school, and it a child one something interesting to do with the pet besides feeding it and cleaning the cage. Many rodents will accept treats and store them in their nest, making it easy to click lots of behavior. Small pets can be trained to do gymnastics and run obstacle courses, to follow a target such as a pencil or a laser pointer, to carry small objects and put them in bottles or boxes, to jump or run through hoops and tunnels on cue. And why draw the line at mammals? One can easily train a fish—as long as it is healthy and eating vigor-ously—to swim through a hoop or jump to a target, using favorite foods as the treat and a flashlight blink as the click.

Lions and tigers and bears...

Some zoos were borrowing marine mammal techniques for use with other animals as early as the 1970's, but as the gen-eral public became familiar with clicker training so did more and more zoos. By 1998 keepers in over a hundred US zoos were using clicker training, even with the largest and most dangerous animals, for routine medical care. Usually the animal is first taught to press its nose against a target, such as a padded pole, for a click and a treat. Using the target, keepers can move animals from one cage to another, or keep them out of the way during cage cleaning.

More important, one keeper can keep the animal stand-ing still at the target while a second keeper provides medical care, a practice the zoos call 'husbandry training.' Keepers now routinely take blood samples and give shots to lions, rhinos, elephants, and polar bears, while the animals volun-tarily stand at their target. One can weigh little rodents and big birds; trim the hooves of giraffes and give badly needed foot care to rhinos; and even teach orangutans and gorillas how to care properly for their babies, all with a clicker and treats.

The animals are spared the stress of the squeeze cage or chemical immobilization, often risky procedures which used to be the only recourse and were saved for emergen-cies. Now every bird and animal in any zoo, including rare and endangered species needed for breeding stock, can have the benefits of stress-free routine treatments, such as immu-nizations, through clicker training.

Working dogs: can they be clicker trained too?

Clicker training is easy and fun for pet owners, and has made life better for many a household dog. But could it ever replace traditional, time-honored methods for training working dogs, such as service dogs that help the handicapped; police dogs that track criminals and help to capture them; and dogs that find hidden drugs and explosives?

Well, yes, if the trainer is willing to experiment. Steve White, a police officer in Seattle, Washington, pioneered the development of clicker training for police work, including tracking and protection, and has taught his techniques to many other trainers in law enforcement. Officer White also teaches clicker training for the search dogs that locate hidden or illegal substances such as drugs, explosives, money, or contraband foodstuffs.

Positive reinforcement, using a marker signal of some sort to provide the dog with accurate information, can facilitate the training of "sniffer" dogs, whatever they are searching for. Experiments are underway in several countries using operant conditioning for dogs trained to locate landmines. One potential advantage is that clicker training can be taught to local handlers, and mine-sniffing skills can be learned by local dogs. At present, however, the methods of organizations that provide dogs for these and other search tasks range from completely traditional to completely modern, with most falling somewhere in between.

While the training of guide dogs for the blind remains as a rule very traditional, programs are more likely to be positively based in the relatively new field of training "service" dogs for the handicapped. These dogs can carry baskets and other objects, pick up things that have been dropped, find and retrieve cell phones, the TV remote, and other objects, turn lights and appliances on and off, close and open doors, provide steadying support for walking or standing, and pull wheelchairs. Many of these programs, such as Dogs for the Disabled in Great Britain, rely on positive reinforcement rather than conventional correction-based training. Nina Bonderenko, director of Canine Partners for Independence, also in Great Britain, develops service dogs entirely by clicker training. Volunteers clicker train puppies from eight weeks on, for housebreaking and general good manners. When young dogs graduate into the formal training program, they already know over fifty verbal cues and useful behaviors.

Formal training of service dog skills is also shaped with the clicker. Recipients of the trained dogs learn clicker training too, so that they can go on adding to the dog's repertoire on their own. Just as important, a person with a physical disability usually finds it much easier to maintain the dog's behavioral skills by clicking for the right behavior than by the more traditional system of physically correcting erroneous behavior.

How about people?

Of course positive reinforcement works with people, and a marker signal is much more accurate than a spoken word in catching behavior as it happens, even in human beings. However, I suppose one might feel a little conspicuous clicking a child for behaving nicely on the bus, say, or in a restaurant. In fact, it's the thought that counts; once you get used to thinking the clicker way, you are much more likely to notice good behavior and reward it, instead of giving your attention to a child only when it's doing something wrong.

Where clicker training actually works brilliantly in humans is in the teaching of physical skills involving timing. Gymnastics is a sport in which things happen much faster than a coach can speak; but a click for correct movement reaches the gymnast even in mid-air. People are experimenting with clickers in training singers, in language training, and in teaching complex multi-part tasks such as flying a plane. And retention of skills learned in this way is permanent. For example a friend of mine used the clicker to teach his three-year-old daughter to balance on ice skates; six months later, when skating season came around again, she balanced perfectly from the start.

The laws of learning are scientific laws; that's what makes clicker training work. The technology of clicker training, however—the rules we are discovering for ourselves as we clicker train—are not yet fully understood. Does the clicker get faster results than a spoken word? Yes, we know that empirically, from all the reports by people who have taught classes with words as the marker signal, and then taught the same material using the clicker. They find that both dogs and their owners learn the new behaviors in roughly half the time, with clicks; we can say it happens, but we certainly can't yet say why. Then there are side effects of clicker training, such as the visible excitement and elation in the learner, and possible changes in brain activity and blood chemistry. Many young scientists, especially a group of students and faculty at the University of North Texas, are researching these and other questions raised by clicker training; some day we will have answers that we don't have now.

What we do have, now, are the observations of thousands of people who have become clicker trainers—and who have noticed and reported results, not just in the learners they are training, but in themselves. Here's what people seem to be finding:

Whether you are working with a pet, a person, or a corporation, the concepts that make up the tools in the clicker trainer's kit lead you to respond in new ways. You become used to the idea that behavior needs to be built in small pieces, not in big chunks. You stop expecting too much, too soon, and just look for what you can reinforce—and so in fact you get more results, and faster.

If you see a behavior you don't like, being a clicker trainer, instead of rushing in to prevent it or stop it, you take it as a training opportunity: what is it this learner needs to know? What's missing? What can I add that would replace this bad behavior? You don't try to push people around with threats and anger, if that was once your style, or with nagging and complaint; you have a better way—reinforcing what you like, instead of attacking what you don't. Some people talk about 'internalizing' the clicker training. Other people just say that once you've crossed over, you can't go back. Again, there are lots of unanswered questions; but any clicker trainer will tell you that this new technology is more than just a training shortcut. The experience can lead to less stress and more fun in life in general—for you, for your pets, and maybe for everyone else around you, too. Click!

–Karen Pryor

6

Resources
Where to Get Help and Learn More

Clicker training resources and gear

www.clickertraining.com

Karen Pryor's clicker training headquarters is a home for the growing international clicker community. Learn from training news and articles, find a clicker trainer in your area, receive email newsletters, join the Clicker Honor Roll (recent wins and achievements by clicker trained dogs, cats, horses), follow links to other clicker Web sites, and locate discussion groups for trainers working with dogs, cats, horses, or birds. Visit the site for secure online ordering of clicker books, videos, and gear, or call 1-800-47 CLICK (781-398-0754).

www.dogwise.com

Large online catalog for dog books and supplies. Gentle Leaders (dog head halters to reduce pulling) in various sizes and colors.

Clicker training gear
- Clicker Fun Packs (Retractable Clicker, Clip-On Clicker, Three Pack). Sturdy replacement clickers and collectible Clicker Fun cards, each with a trick you can teach your dog in 10 easy steps $7.95
- Target Stick. Folding aluminum target stick makes training a snap $16.95

Getting Started Clicker Kits

Includes everything you need to get going: Getting Started: Clicker Training Kits include book, pocket guide, two clickers, and treats.

Getting Started: Clicker Training for Dogs KIT $16.95
Getting Started: Clicker Training for Cats KIT $16.95
Getting Started: Clicker Training for Horses KIT $16.95
Coming soon!
Getting Started: Clicker Training for Birds KIT $16.95

Clicker training books

Clicking with Your Dog, Step-by-Step in Pictures, by Peggy Tillman. Teach yourself clicker training with clear step-by-step pictures! This new book makes it easy and fun. More than 100 behaviors, including trouble-free fun for dogs left home alone, housetraining, and cute tricks. Each picture shows you exactly when to click and when to treat. $24.95

Click to Win: Clicker Training for the Show Ring. Karen Pryor's popular articles from the AKC Gazette collected into one full-color volume. You'll learn how to teach any dog of any breed and any age (including puppies just weeks old) how to self-stack and gait like a champion in the conformation ring. Also, chapters on clicking puppies, on manners, and on advanced clicker training $24.95

Clicker Training for Obedience by Morgan Spector shows how and why to use clicker training to develop the obedience skills for competition, service, and companion dogs. Whether you're starting off with a new puppy or headed for the obedience show ring, Spector tells you exactly how to get started and how to achieve success. This large-format, 260 page book is packed with illustrations, and includes a glossary and a resource section. $29.95

Clicker Training for Your Horse by Alexandra Kurland will help you turn the horse you have into the horse of your dreams. This large-format book offers 200 pages of clear, exciting advice and photos on how clicker training helps you communicate with your horse clearly and without force $29.95

Don't Shoot the Dog! The New Art of Teaching and Training, new revised edition by Karen Pryor. The bible on reinforcement training, not just for pets but in the classroom, in sports, and in life. An American Psychology Association award winner. More than 300,000 copies in print. $13.95

Lads Before the Wind, Expanded Edition. Karen Pryor's classic memoir of training dolphins in Hawaii, when clicker training was done with a whistle and a bucket of fish $16.95

Clicker training videos

Puppy Love

Karen Pryor introduces clicker training as the pet-friendly way to raise a great family dog. Now everyone in the family can help the new puppy learn where to sleep, the right place to "go," how to sit to greet guests instead of jumping up, how to come when called, to do cute tricks, to walk on a loose leash, and more. Puppy Love is full of valuable advice for newly adopted shelter dogs, too. (Video comes with free clicker.)

VHS (US format). $24.95
PAL (European format) $34.95

Clicker Magic!

Twenty real-time clicker training sessions with dogs, puppies, a mule, a fish, and a cat demonstrate the fundamentals, the timing, and the potential of clicker training.

VHS (US format) $39.95
PAL (European format) $49.95

Shipping and handling additional on all orders.
To order, call 1-800-47CLICK or visit www.clickertraining.com

Clicker Discussion Lists

www.clickersolutions.com
A beginner-friendly Web site and discussion list for owners of dogs and other animals run by Melissa Alexander, author of *Clicker Solutions: The Clicker Training Answer Book* (fall 2002).

www.click-l.com
The original clicker list and Web site with a huge membership and close monitoring to stay on subject.

The Well Mannered Dog: www.shirleychong.com
This Web site, run by clicker trainer and teacher Shirley Chong, has a huge variety of tips and ideas on agility, obedience, tricks, and lots more.

www.bestbehavior.net
This helpful Web site is run by clicker trainer Morgan Spector, author of *Clicker Training for Obedience*.